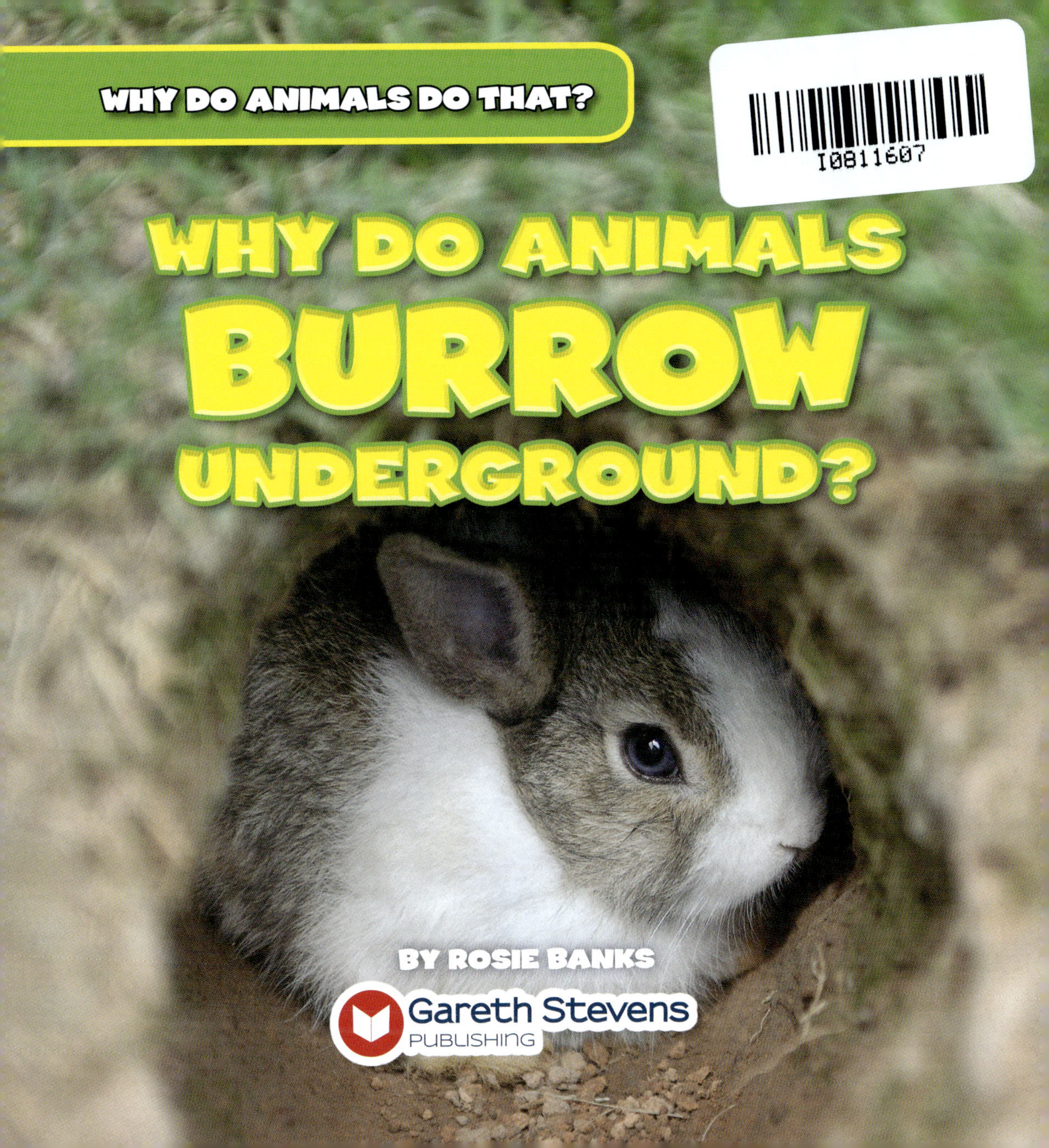

WHY DO ANIMALS DO THAT?

WHY DO ANIMALS BURROW UNDERGROUND?

BY ROSIE BANKS

Gareth Stevens PUBLISHING

Please visit our website, www.garethstevens.com. For a free color catalog of all our high-quality books, call toll free 1-800-542-2595 or fax 1-877-542-2596.

Library of Congress Cataloging-in-Publication Data

Names: Banks, Rosie, 1978- author.
Title: Why do animals burrow underground? / Rosie Banks.
Description: New York : Gareth Stevens Publishing, [2024] | Series: Why do animals do that? | Includes bibliographical references and index.
Identifiers: LCCN 2022045691 (print) | LCCN 2022045692 (ebook) | ISBN 9781538285091 (library binding) | ISBN 9781538285084 (paperback) | ISBN 9781538285107 (ebook)
Subjects: LCSH: Animal burrowing–Juvenile literature.
Classification: LCC QL756.15 .B36 2023 (print) | LCC QL756.15 (ebook) | DDC 591.56/48–dc23/eng/20221003
LC record available at https://lccn.loc.gov/2022045691
LC ebook record available at https://lccn.loc.gov/2022045692

Published in 2024 by
Gareth Stevens Publishing
2544 Clinton Street
Buffalo, NY 14224

Designer: Claire Wrazin
Editor: Therese Shea

Photo credits: Cover ANUCHID.L/Shuttertstock.com; Series Art (texture) Anna Timoshenko/Shutterstock.com; p. 5 Yashkin Ilya/Shutterstock.com; p. 7 Fenykep Formatumok/Shutterstock.com; p. 9 Christian Schwier/Shutterstock.com; p. 9 (inset) juefraphoto/Shutterstock.com; p. 11 Evoken/Shutterstock.com; p. 13 File:Tipton_kangaroo_rat_8046-5_1991.jpg/Wikimedia Commons; p. 15 Bolshakov_Ilya/Shutterstock.com; p. 17 Antonio Volb/Shutterstock.com; p. 19 Geoffrey Kuchera/Shutterstock.com; p. 21 File:Oryctodromeus_cubicularis_-_Museum_of_the_Rockies.jpg/Wikimedia Commons.

Printed in the United States of America

CPSIA compliance information: Batch #CSGS24: For further information contact Gareth Stevens, New York, New York at 1-800-542-2595.

CONTENTS

Boldface words appear in the glossary.

Why Go Underground?

Look at that hole! What animal dug it? Why did they dig it? The holes and tunnels animals dig under the ground are called burrows. They can be deep or close to the **surface**. Animals dig burrows for many reasons.

Home in a Hole

Prairie dogs are **rodents**. They live in groups called colonies. Their burrows have many tunnels and rooms. Some rooms are for sleeping. Some are used like toilets! The biggest burrow found was about 100 miles (161 km) wide and 250 miles (402 km) long.

Eating Underground

Moles spend most of their lives underground. They dig tunnels as they look for food, such as worms and bugs. They dig a bit like they're swimming. You may see mounds of dirt above their burrows. These are called molehills.

Protection from Predators

Many species, or kinds, of rabbits dig burrows. They use their burrows to stay safe from predators. They sleep and raise their young in them. Rabbit burrows are sometimes called warrens. Warrens can be 15 feet (4.6 m) long!

Beating the Heat

Kangaroo rats are rodents that live in some North American deserts. They hide from hot **temperatures** in their burrows. The burrows are cooler than the air above. Kangaroo rats also use their burrows to store food, such as seeds.

Arctic Escape

Arctic ground squirrels hide from the cold in burrows. They dig them on higher ground. They can see far around them when they come out. Arctic ground squirrels **hibernate** in burrows through the winter. Some frogs hibernate in burrows too!

Burrow for the Body

Reptiles are cold-blooded. That means their body temperature changes with their **environment**. Temperatures underground don't change as quickly as on the surface. Reptiles, such as some lizards and snakes, burrow to keep their body a healthy temperature. They may hide in burrows too.

Moving In

Some animals leave their burrows to look for food in another area. Or they may leave a burrow after a predator finds it. Other animals use these burrows. Foxes sometimes take other animals' burrows. They may dig their own too.

Dinos Too!

Scientists discovered that some dinosaurs dug burrows. Dinosaur **fossils** were found in a burrow in Montana in 2007. The dinosaur was named *Oryctodromeus* (or-ik-tow-DROHM-ee-us). That means "burrowing runner." Even some birds and fish use burrows. This **adaptation** helps so many animals stay alive!

WHY ANIMALS BURROW

To stay safe from predators

To guard themselves from heat and cold

To travel from one place to another

Oryctodromeus

To keep their bodies healthy

To hide food underground

To hibernate

To find food underground

GLOSSARY

adaptation: A change in a type of animal that makes it better able to live in its surroundings.

environment: The natural world in which a plant or animal lives.

fossil: The marks or remains of plants and animals that formed over thousands or millions of years.

hibernate: To be in a sleeplike state for a long period of time, usually during winter.

reptile: An animal covered with scales or plates that breathes air, has a backbone, and lays eggs, such as a turtle, snake, lizard, or crocodile.

rodent: A small, furry animal with large front teeth, such as a mouse or rat.

surface: The top layer of a planet. Also, the ground.

temperature: How hot or cold something is.

FOR MORE INFORMATION

BOOKS

Golkar, Golriz. *Meerkats.* Minneapolis, MN: Pop!, 2022.

Ruby, Rex. *Inside a Prairie Dog's Hideaway.* Minneapolis, MN: Bearport Publishing Company, 2023.

WEBSITES

Burrowing Owl
www.audubon.org/field-guide/bird/burrowing-owl
Read about a bird that burrows underground.

Meerkat
kids.nationalgeographic.com/animals/mammals/facts/meerkat
Meerkats sometimes have five burrows at once! Learn more about these African animals.

Publisher's note to educators and parents: Our editors have carefully reviewed these websites to ensure that they are suitable for students. Many websites change frequently, however, and we cannot guarantee that a site's future contents will continue to meet our high standards of quality and educational value. Be advised that students should be closely supervised whenever they access the internet.

INDEX